Erratics

Erratics

photographs by
Charles Katz, Jr.

with essay by
Dr. Kevin R. Pogue

Published by
Sheehan Art Gallery
Whitman College
Walla Walla, Washington 99362
in cooperation with Obscurity Press

LCCN 2002112353
ISBN 1-880269-16-3

First Edition, 2002

In memory of Charles J. Katz (1919-2002),
photographer and geologist

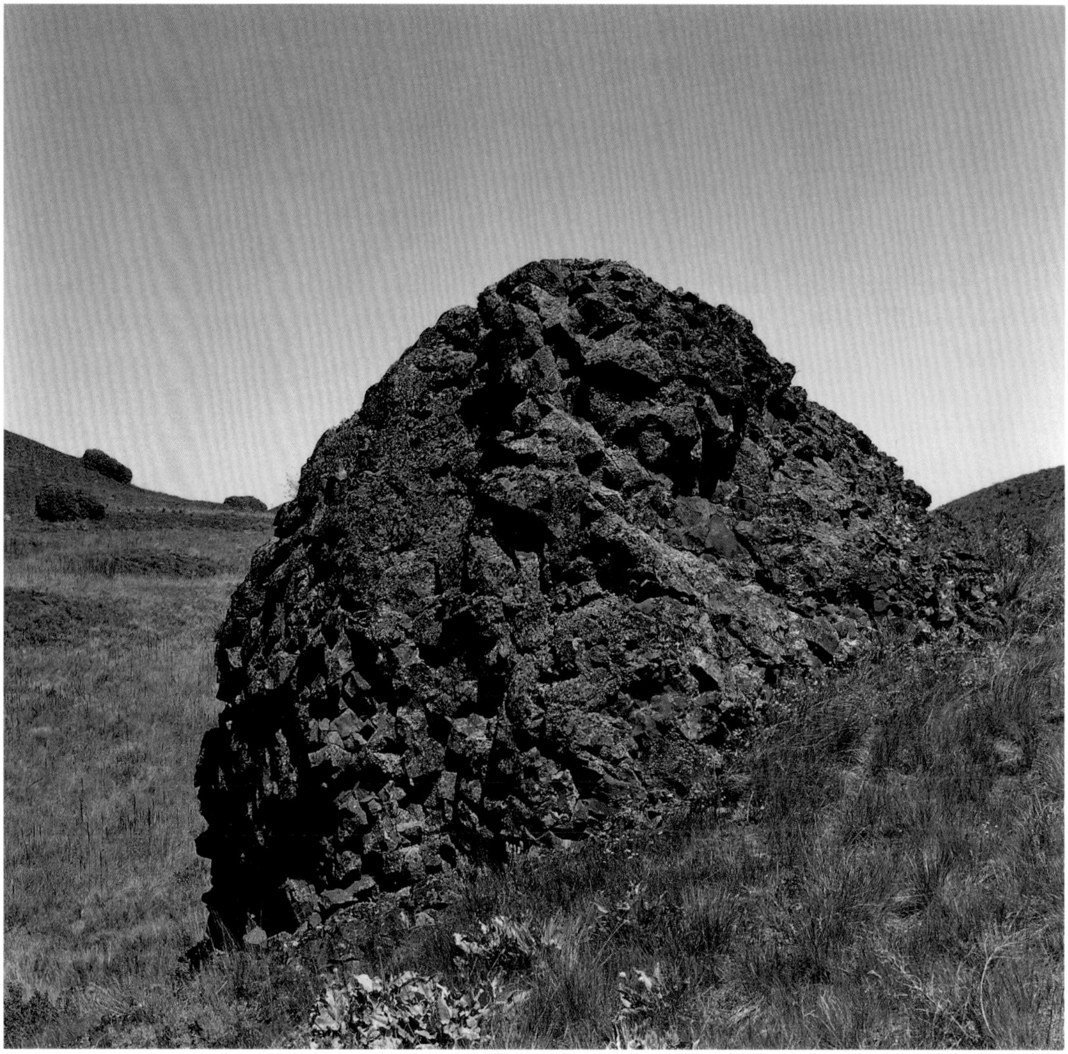

The birds have vanished down the sky.
Now the last cloud drains away.

We sit together, the mountain and me,
until only the mountain remains.

—Li Po, "Zazen on Ching-t'ing Mountain"

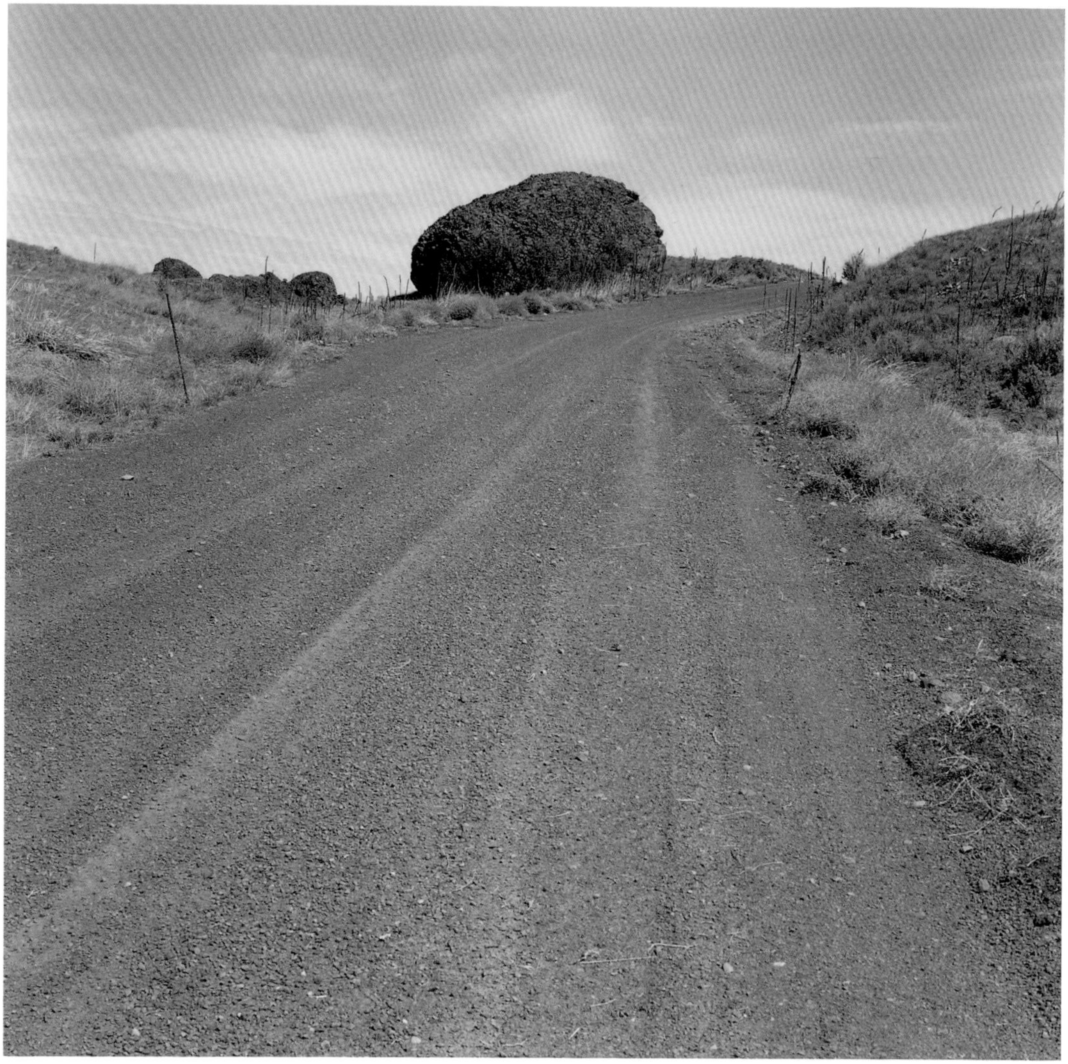

Monolithic Erratics of the Withrow Moraine

The monolithic glacial erratic boulders in these photographs by Chuck Katz are a prominent feature of the Withrow Moraine on the Waterville Plateau in north-central Washington state. The geologic history of these "haystack rocks," as they are known in the area, is one of floods—floods of lava, of water, and of ice.

FLOOD BASALTS

The haystack rocks and the bedrock on which they lie are made of basalt, a dark, fine-grained volcanic rock formed between six million and 16 million years ago when excess heat partially melted the earth's mantle to produce magma, which erupted through fractures in the earth's crust. These eruptions, centered in southeastern Washington, produced a series of massive lava floods that flowed northwest across central Washington, filling valleys, displacing streams and rivers, and ultimately emplacing more than 35,000 cubic miles of basalt over an area of 60,000 square miles.

Among those rivers displaced by the volcanic floods was the ancestral Columbia, diverted to the northern and western margin of the thick basalt pile where—when the eruptions ceased—the river and its tributaries began to incise their channels. Flowing along the fringes of the basaltic plain, major tributaries such as the Methow and Okanogan rivers removed most of the relatively thin layer of basalt from their valleys, while the much larger Columbia River incised a canyon into the margins of the main basaltic mass. That canyon now exposes a cross-section of the lava flows that form the bedrock of much of central Washington.

The cliffs that line the river display numerous geologic structures produced primarily by the basalt's cooling. As each successive basaltic lava flow cooled and contracted, it developed a network of cracks called joints. The lower, more insulated part of each lava flow developed regularly spaced joints, creating large polygonal columns that form spectacular palisades along the Columbia River. Known as the colonnade, this section of each flow was overlaid by the entablature, a zone featuring closely spaced joints that outline much smaller columns in a chaotic arrangement. Finally, once the lava had cooled completely, tectonic stresses in the crust created widely spaced regional joints that cut through the entire series of lava flows.

ICE SHEETS AND *JÖKULHLAUPS*

Approximately two million years ago, the earth's climate cooled, and the most recent ice age began. Several times during this period, an ice sheet formed in Canada and advanced southward before retreating, with each major advance known as a glaciation. During the most recent glaciation, which ended about 10,000 years ago, a lobe of the ice sheet moved south into northern Idaho and Montana and blocked the Clark Fork River at Cabinet Gorge, just east of Lake Pend Oreille. The river ponded behind this ice dam, creating a lake—Glacial Lake Missoula—that eventually inundated most of the major valleys of western Montana. The lake was as much as 2,000 feet deep and contained over 500 cubic miles of water; the site of the city of Missoula was under 950 feet of water.

Ultimately, the water reached a depth sufficient to float the ice dam, and the lake suddenly drained, releasing what geologists call a *jökulhlaup*, an Icelandic term meaning "glacial outburst flood." Tunneling beneath the ice at a flow rate that eventually exceeded 350 million cubic feet per second, the water eroded and enlarged its pathway, surging southwestward across northern Idaho and east-central Washington, into the valleys of the Columbia River and its tributaries. In many cases, the volume of water overwhelmed the existing river valleys, and the floods spilled out across the adjoining landscape, carving new valleys in the basalt bedrock that are now called "coulees." This sequence of events was repeated as many as 100 times between 15,300 and 12,700 years ago, the continental ice sheet continuing to move south after each dam was destroyed, reestablishing the blockage.

Some of Glacial Lake Missoula's earlier floods filled the existing channel of the Columbia River and overflowed to the south, a process that repeatedly inundated the Waterville Plateau and eroded numerous small coulees and one major one, Moses Coulee. In later years, however, another huge tongue of ice invading southward down

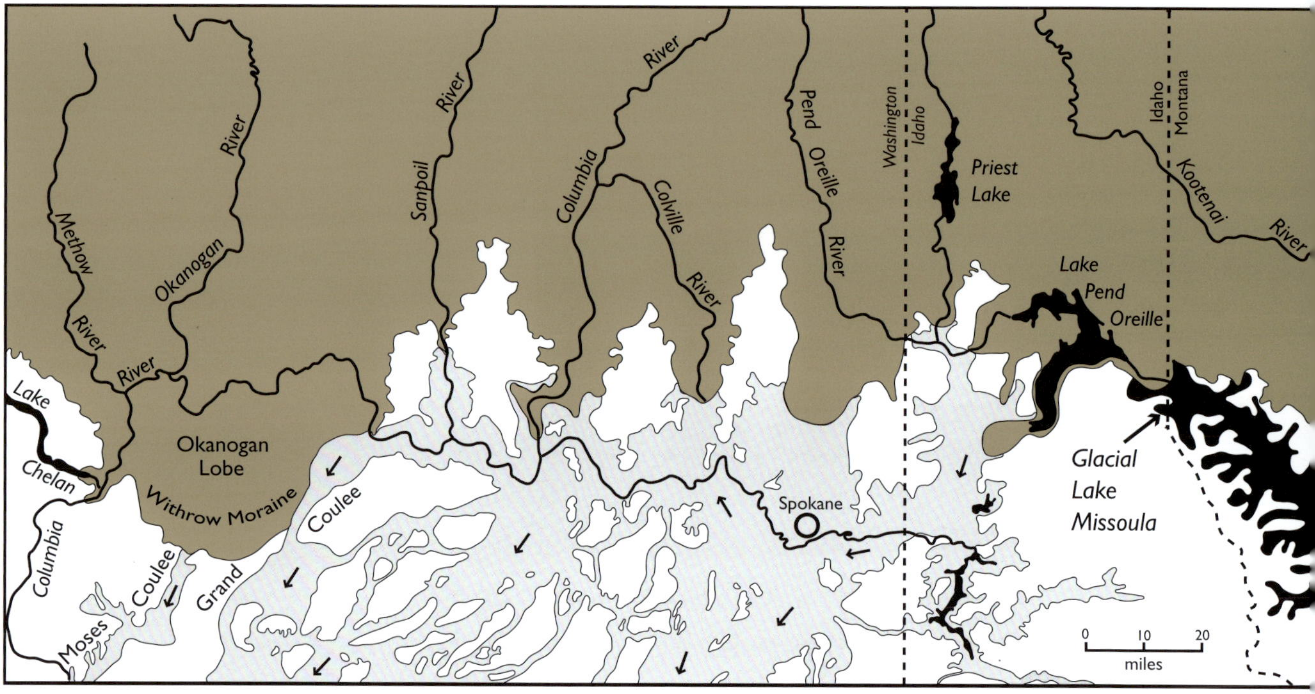

maximum area covered by ice during Pleistocene glaciation

← areas inundated by Missoula floods showing flow direction

the valley of the Okanogan River eventually reached the Columbia near the site of Grand Coulee Dam, where it formed a second ice dam. This ice blocked the river, creating Glacial Lake Columbia in approximately the same location as the modern reservoir called Roosevelt Lake. Unlike Glacial Lake Missoula, Glacial Lake Columbia had an outlet. When the impounded waters of the Columbia rose high enough, they spilled over the valley wall to the south, eroding Grand Coulee. The Okanogan ice lobe eventually extended another 30 miles south onto the Waterville Plateau, where its southernmost extent is marked by a mile-wide terminal moraine. At its maximum size, the ice completely filled the Columbia's 2,000-foot-deep canyon and rose 1,000 to 2,000 feet above the canyon rim. This enormous frozen mass diverted the entire volume of Glacial Lake Missoula floods southward out of the Columbia's channel, and the waters scoured their way across central Washington before rejoining the Columbia River farther south.

ICE AND WATER MEET BASALT—THE HAYSTACK ROCKS

Pushing across the Columbia River, the Okanogan ice lobe encountered the jointed basalt cliffs that line the river gorge and the coulees of the Waterville Plateau. As the advancing ice enveloped the cliffs, meltwater penetrated the joint systems and refroze, allowing the glacier to pluck and capture large chunks of basalt. The ice easily penetrated the long, continuous joints of the colonnade, which was readily dismembered into small chunks of rock the width of individual columns and carried away. The shorter, discontinuous joints of the entablature, on the other hand, were not so easily exploited, and the glacier instead quarried out the entablature in large blocks along the regional joints, spaced at 30- to 100-foot intervals. Some of these boulders were transported to the snout of the glacier; as the ice melted, these rocks were deposited in the terminal moraine. Other large boulders made it only partway to the glacier's front before the ice stopped moving. When the ice melted, the boulders were deposited in distinct trains, oriented parallel to the flow direction of the glacier. These lines of boulders can sometimes be traced upstream to the cliffs from which they were derived.

The haystack rocks are the largest ice-transported boulders within the blanket of glacial debris that covers the Waterville Plateau. Their size stems from the spacing of the regional joints that formed their boundaries within the entablature, in addition to the rocks' proximity to the parent cliffs: Boulders transported farther by the glacier were more likely to be dismembered, so the largest rocks either were quite big to begin with or lie closer to their sources. After they were deposited by the ice, the haystack rocks were attacked by weathering and erosion. Most are now somewhat rounded and ringed by an accumulation of talus derived from the action of frost-wedging along small joint planes. Some have already disintegrated completely into piles of small rocks.

—Dr. Kevin R. Pogue
Associate Professor of Geology, Whitman College

Section I

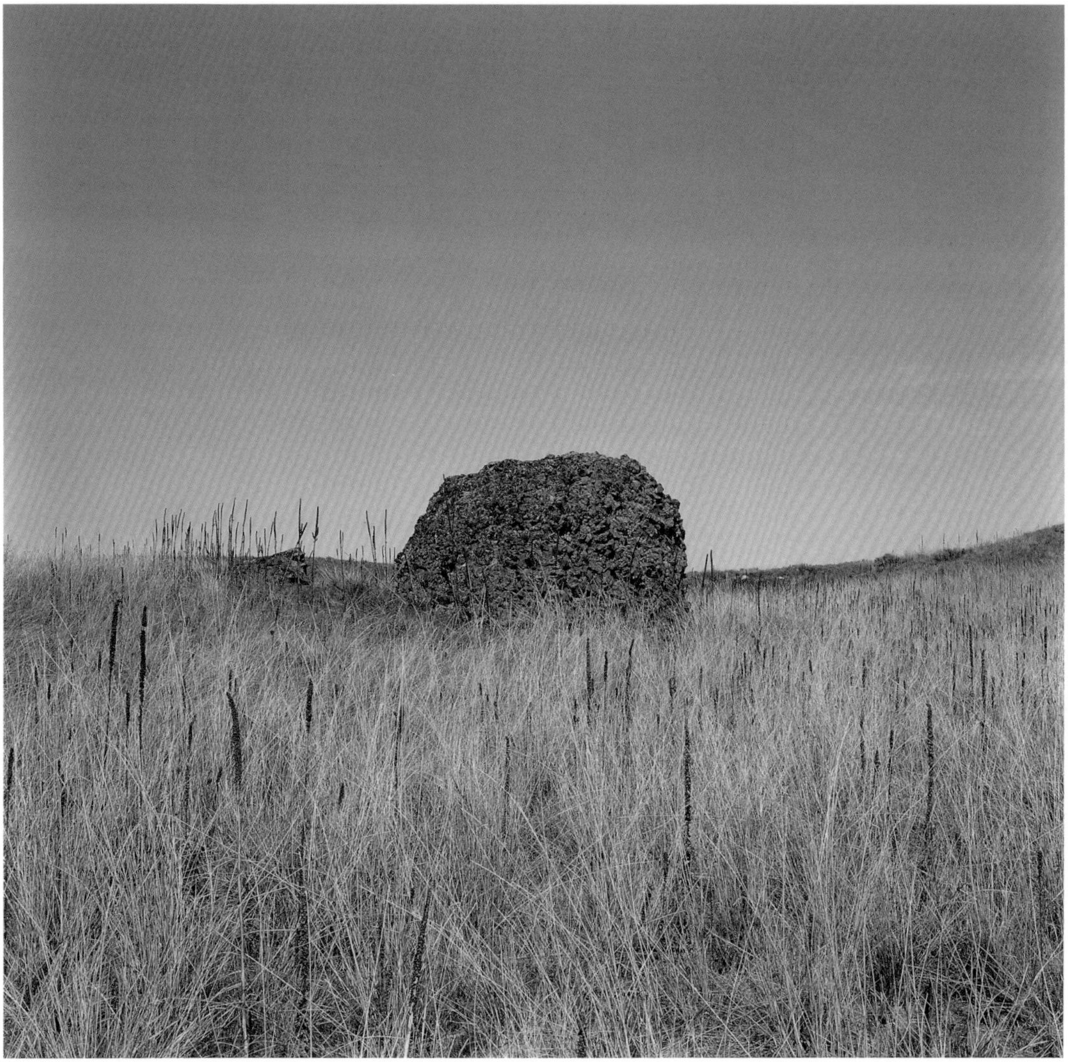

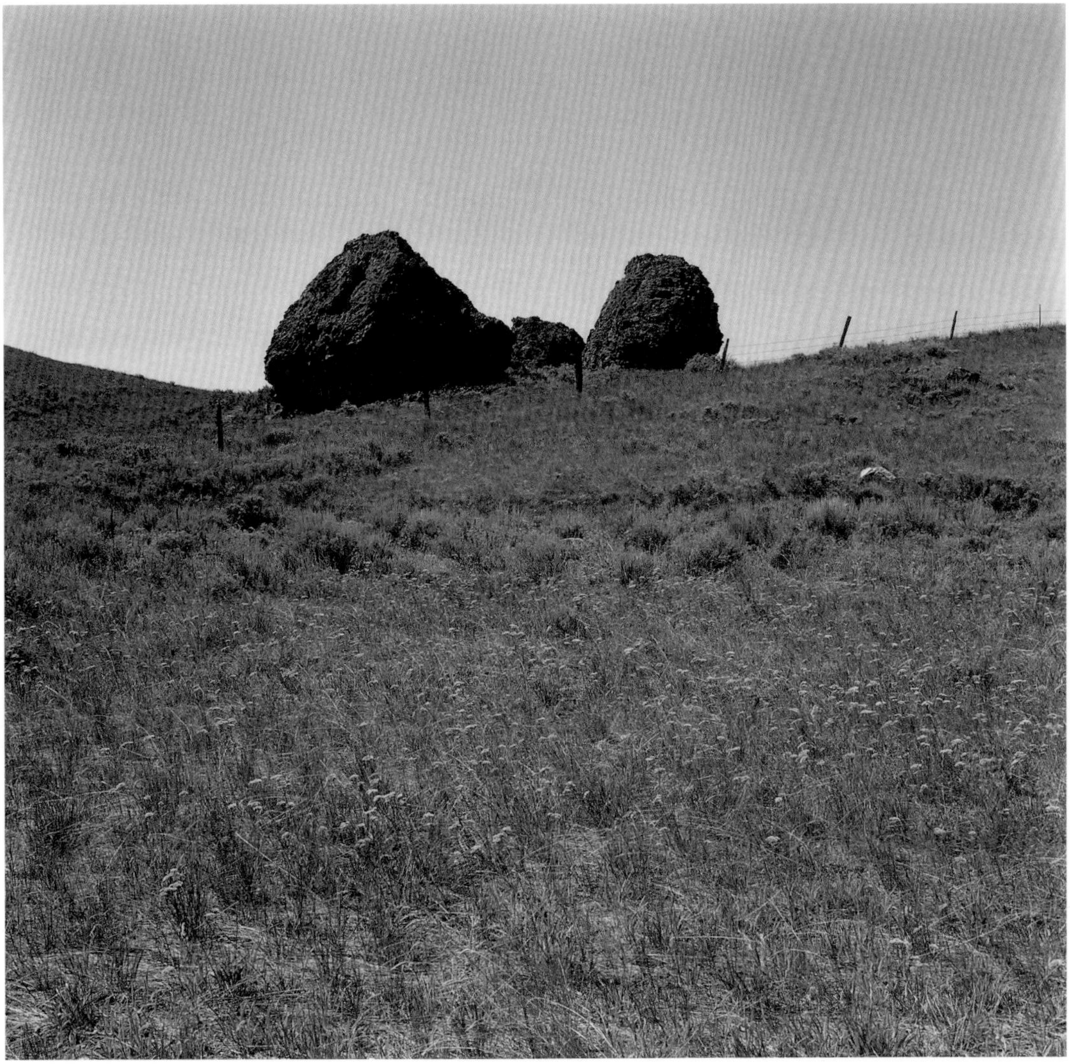

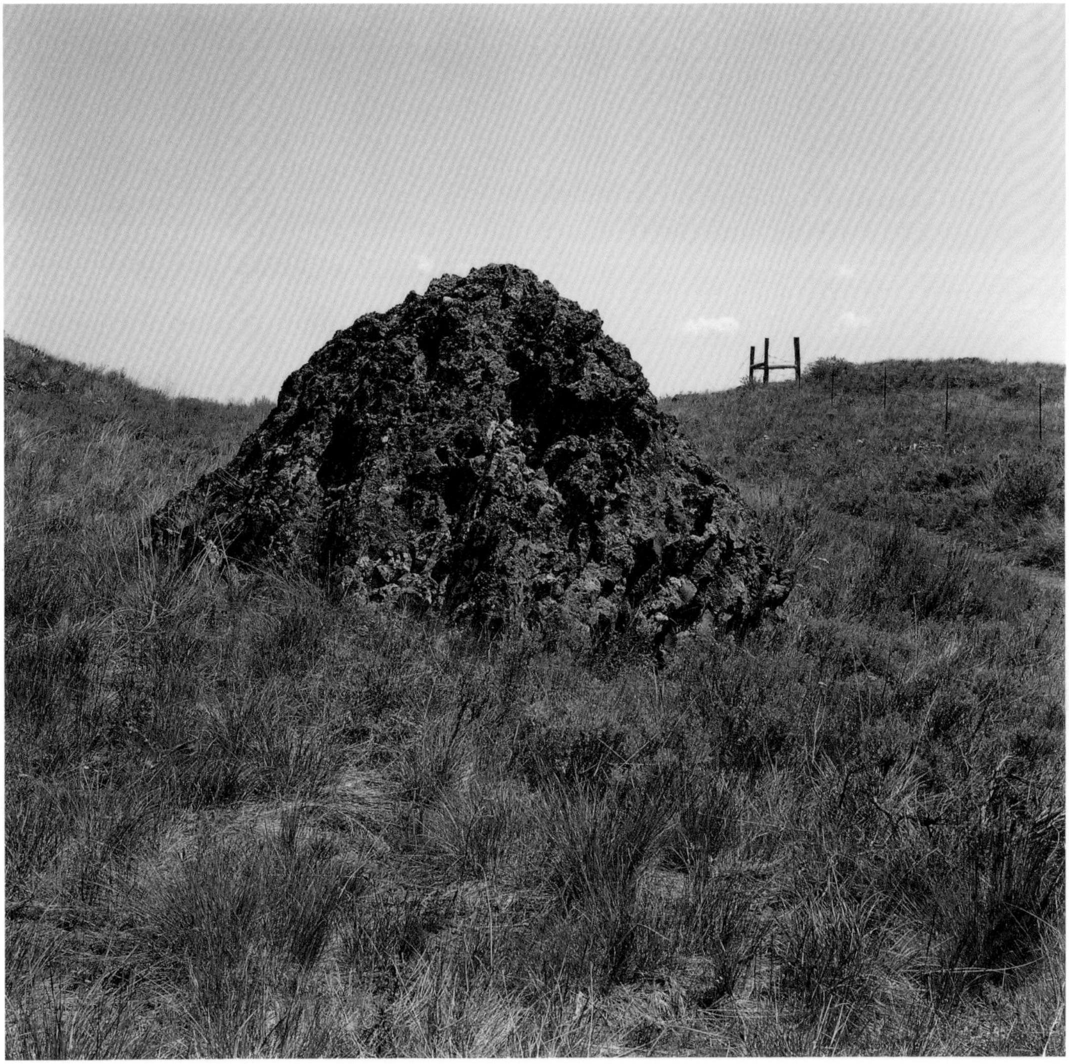

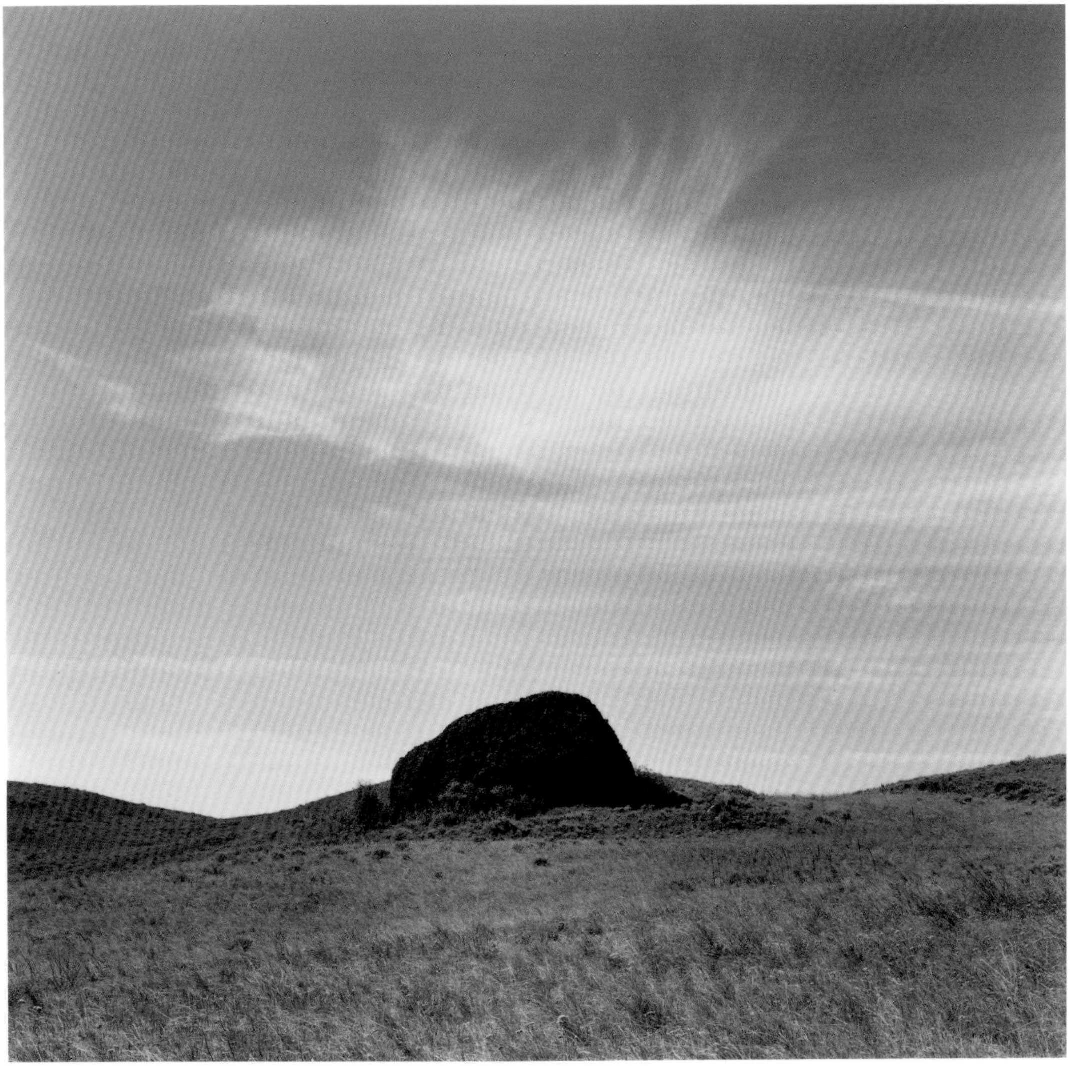

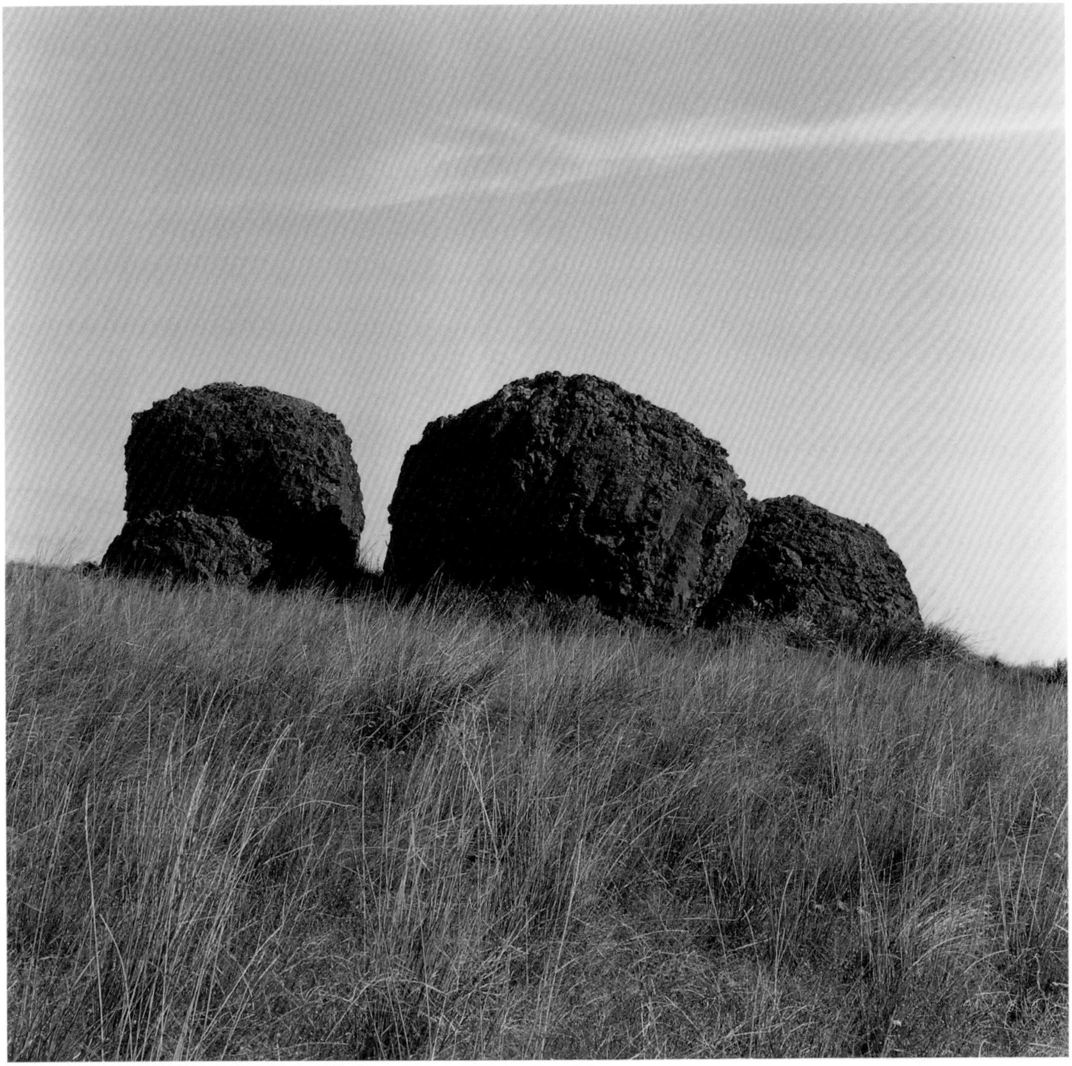

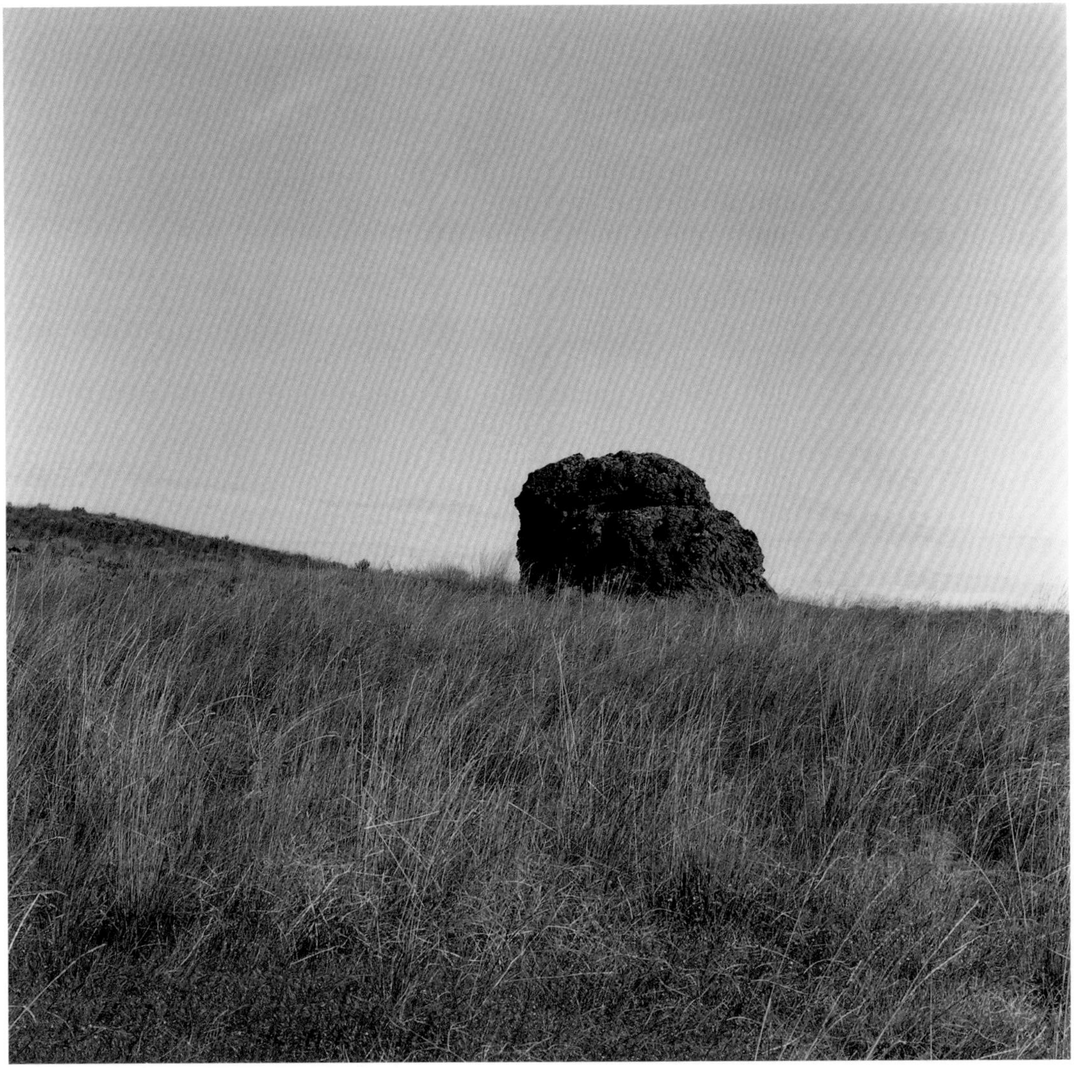

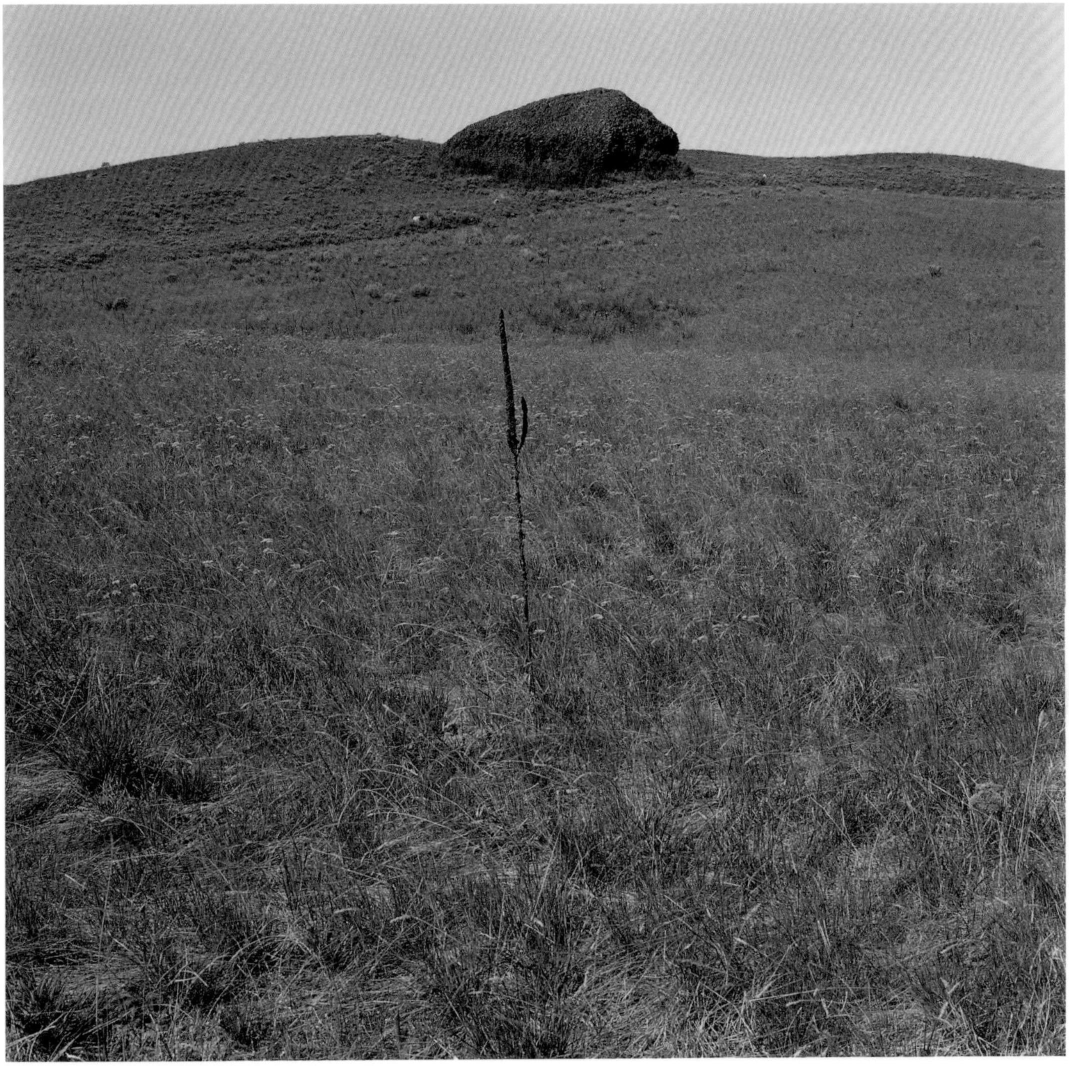

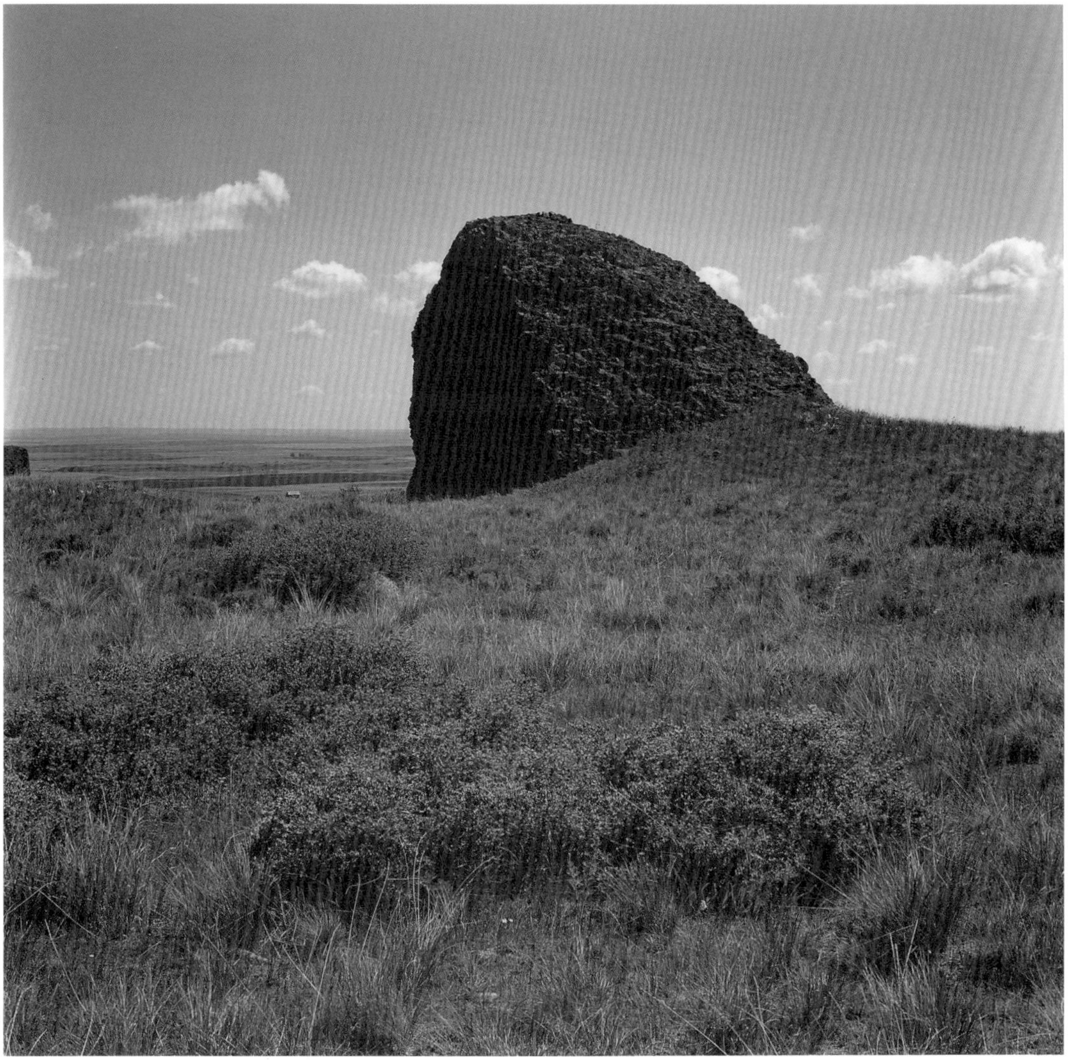

Section II

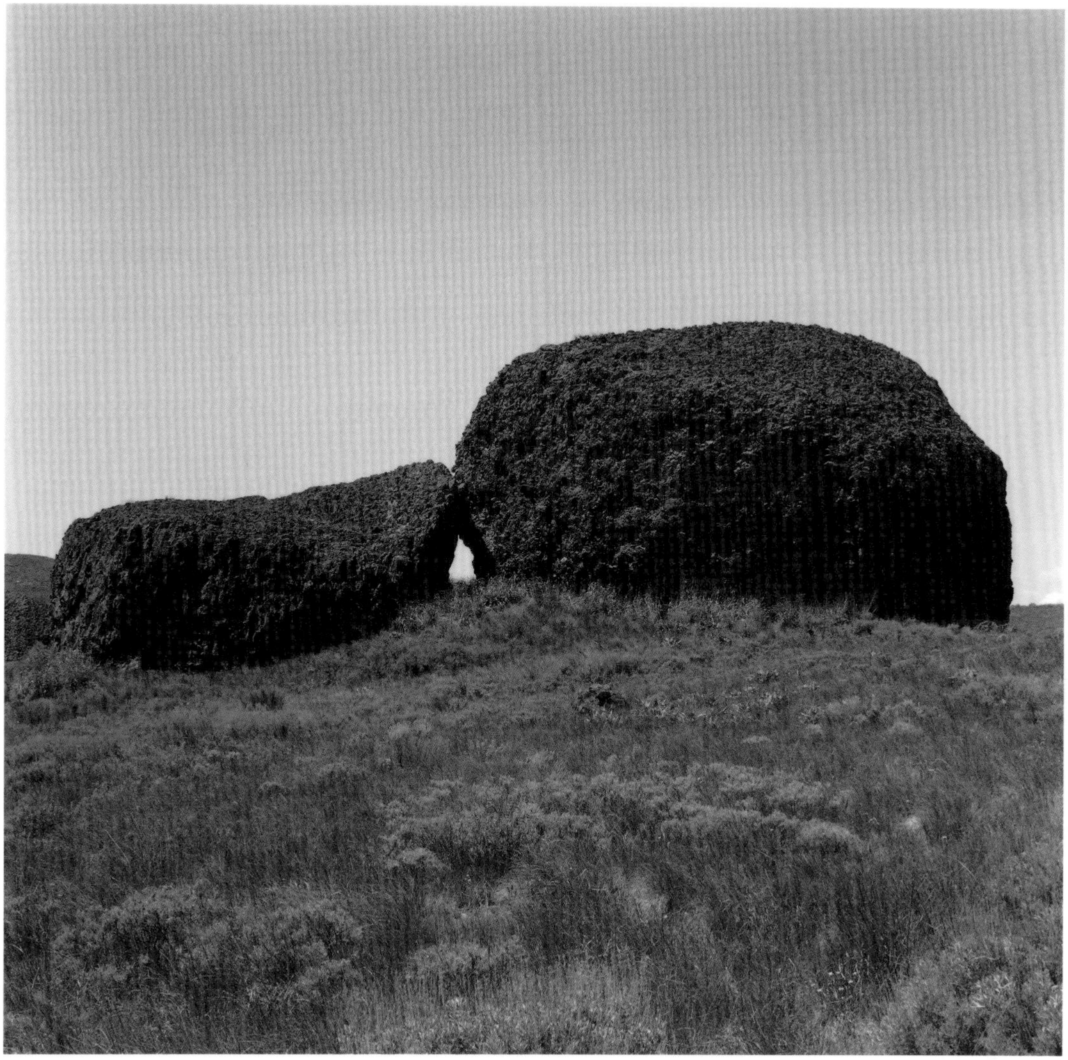

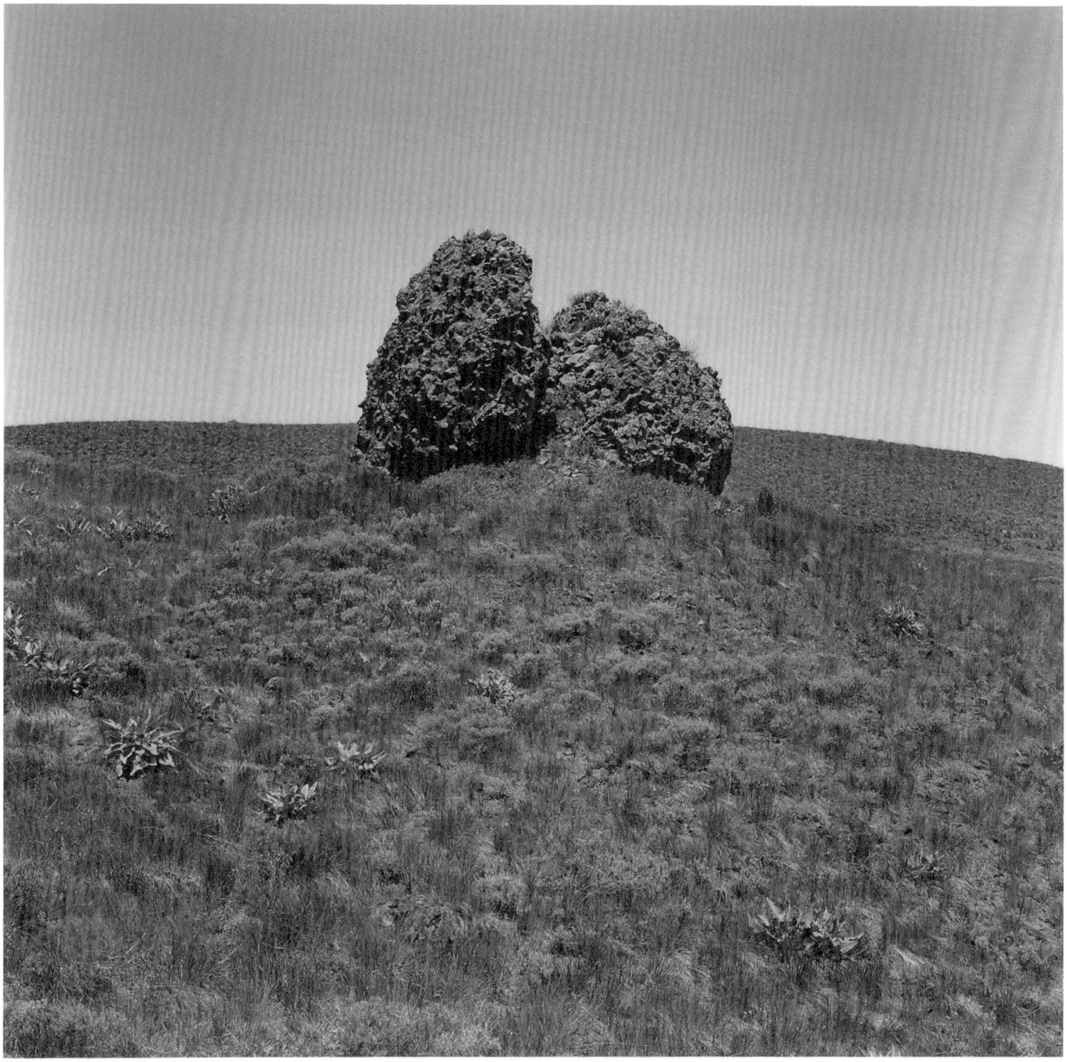

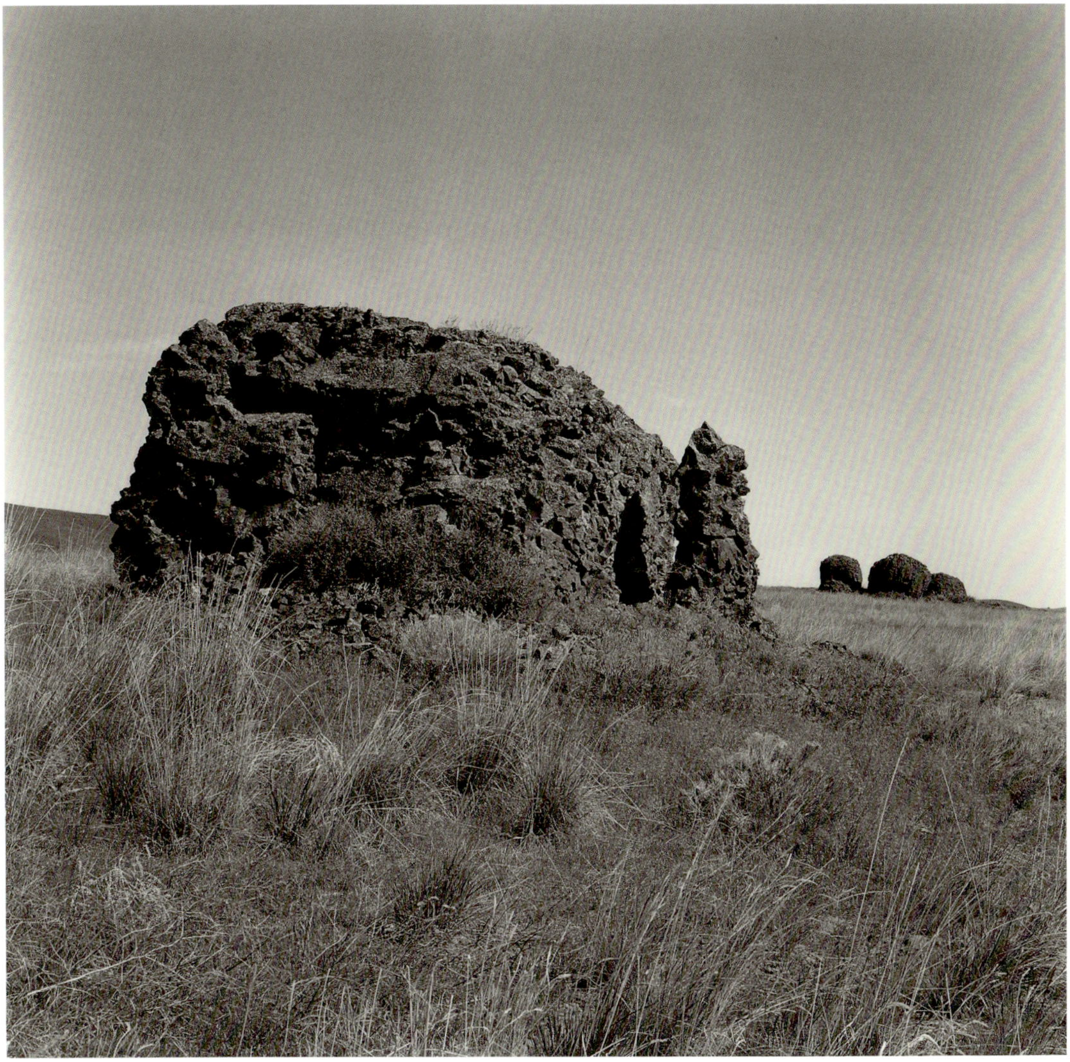

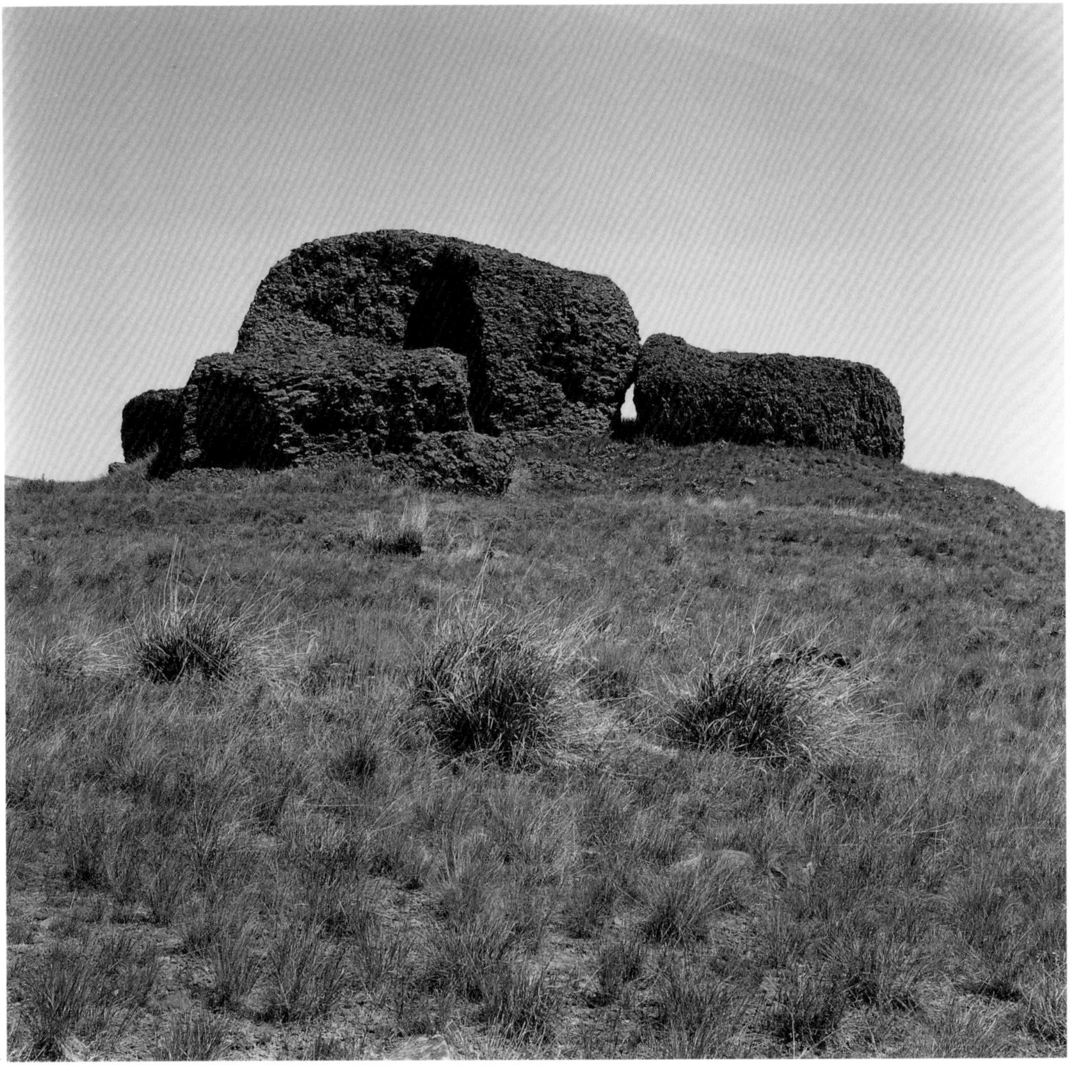

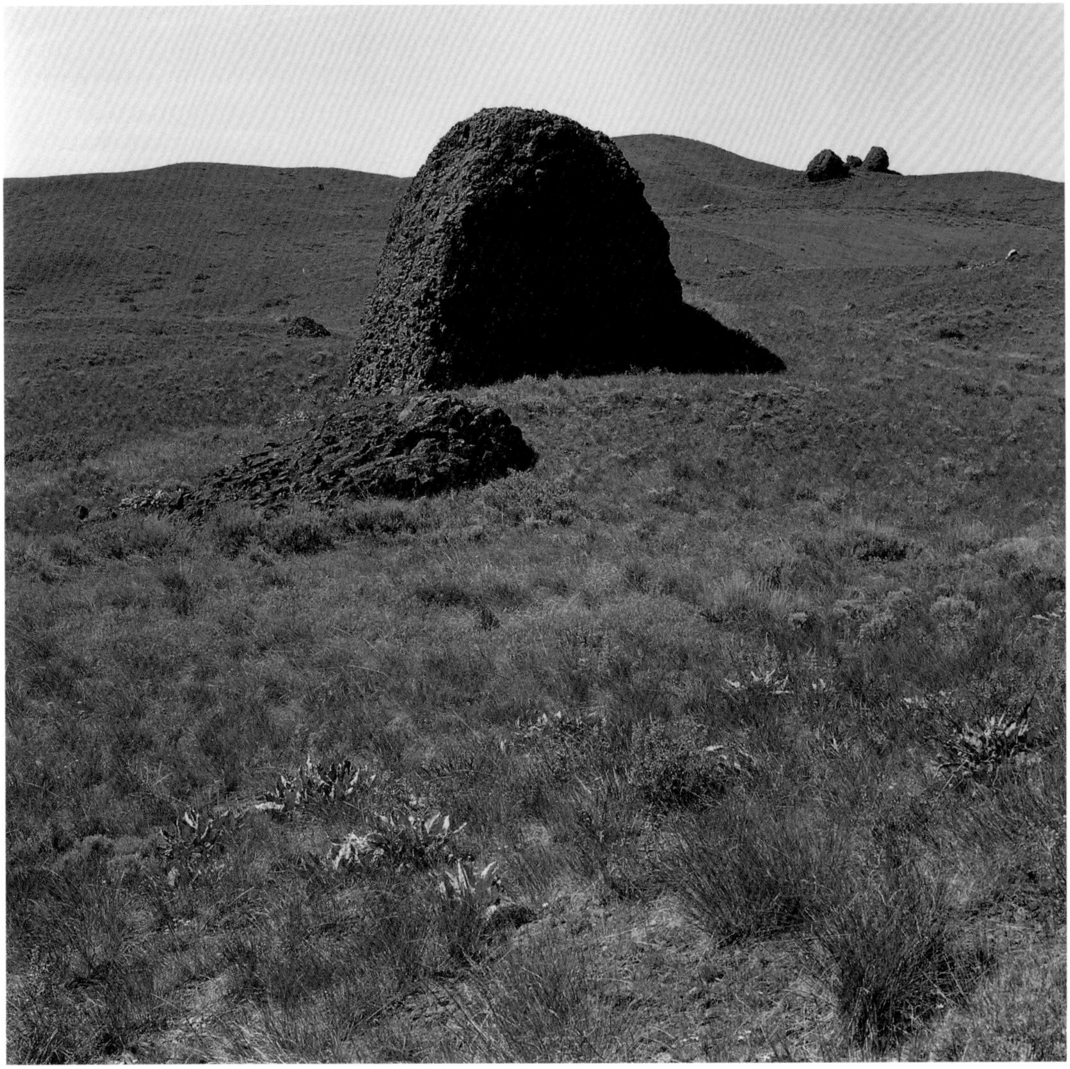

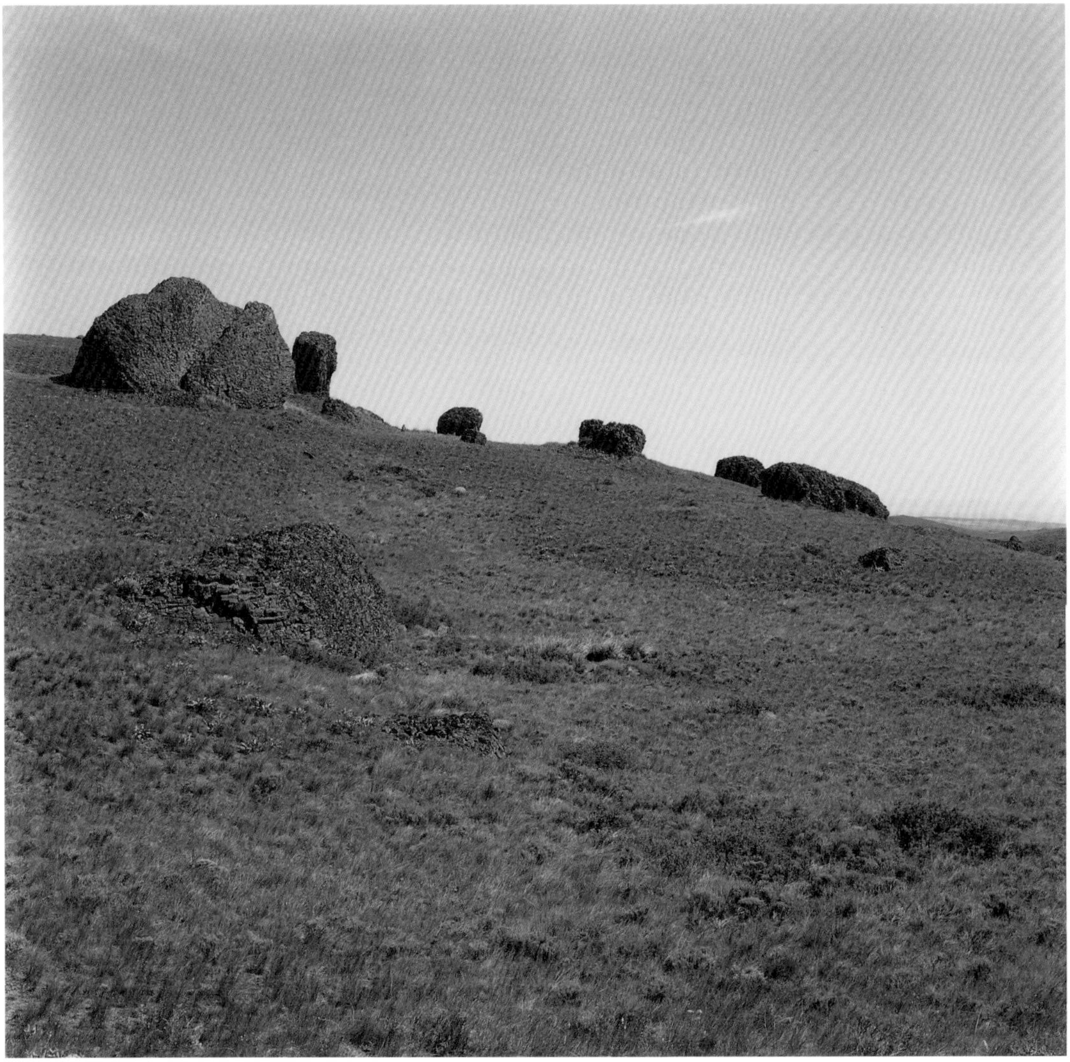

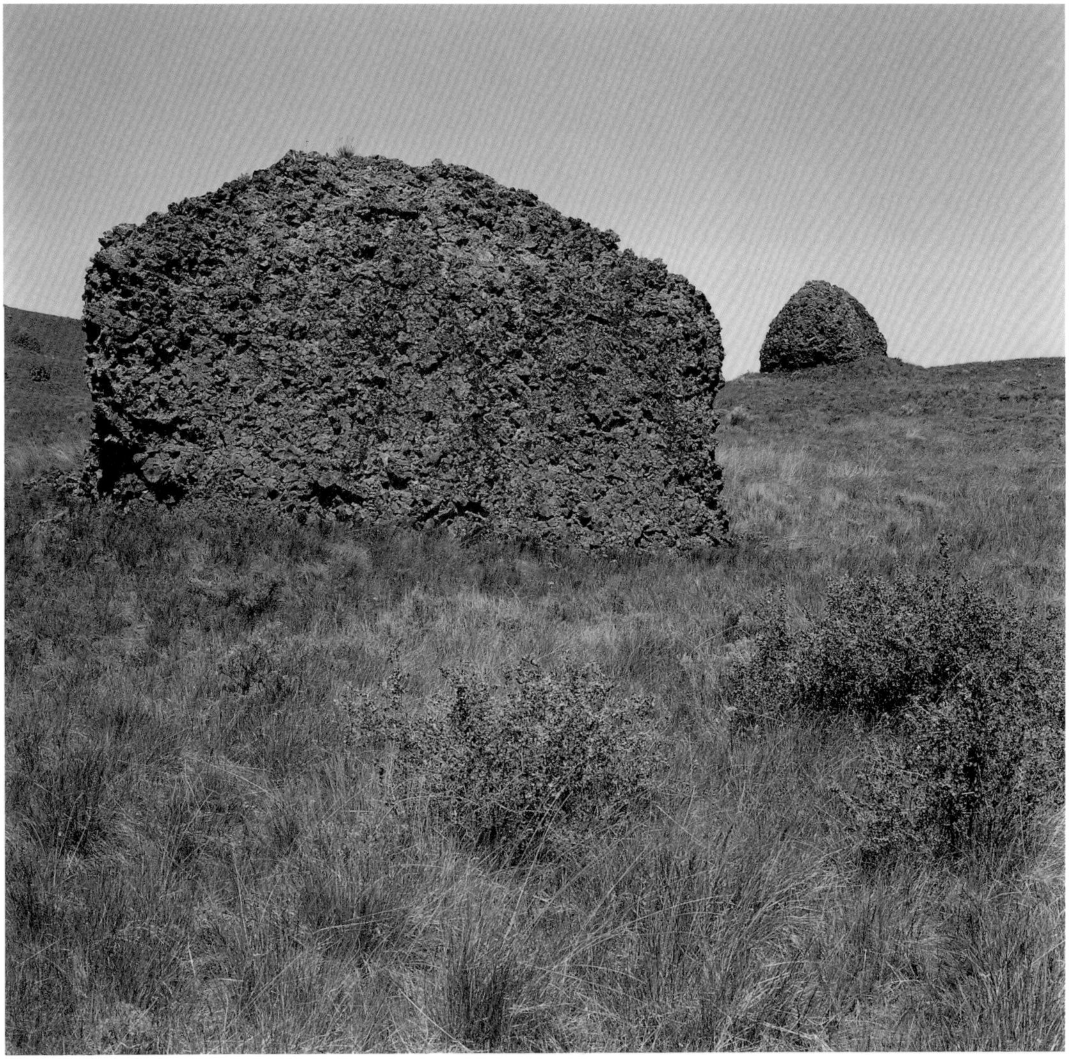

Section III

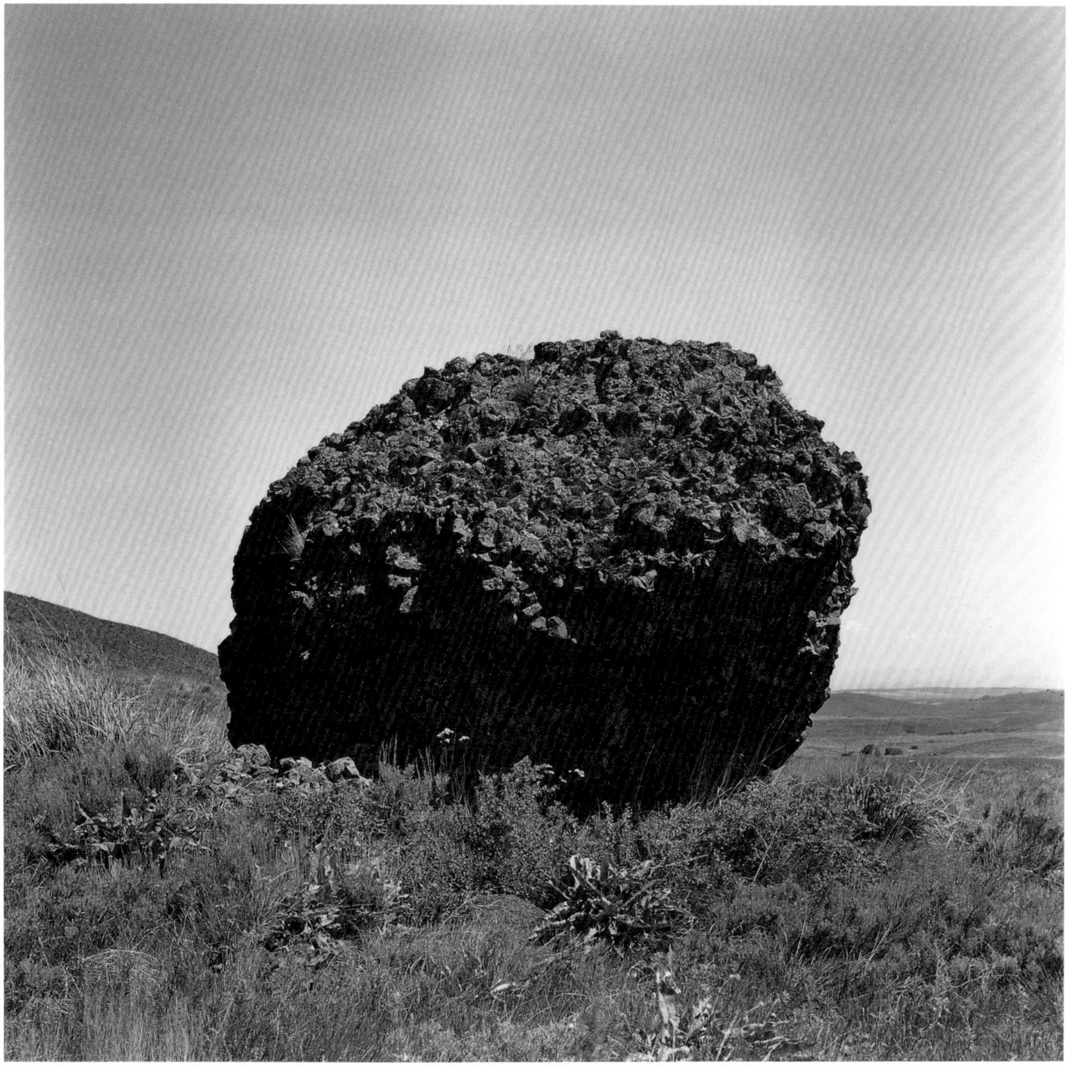

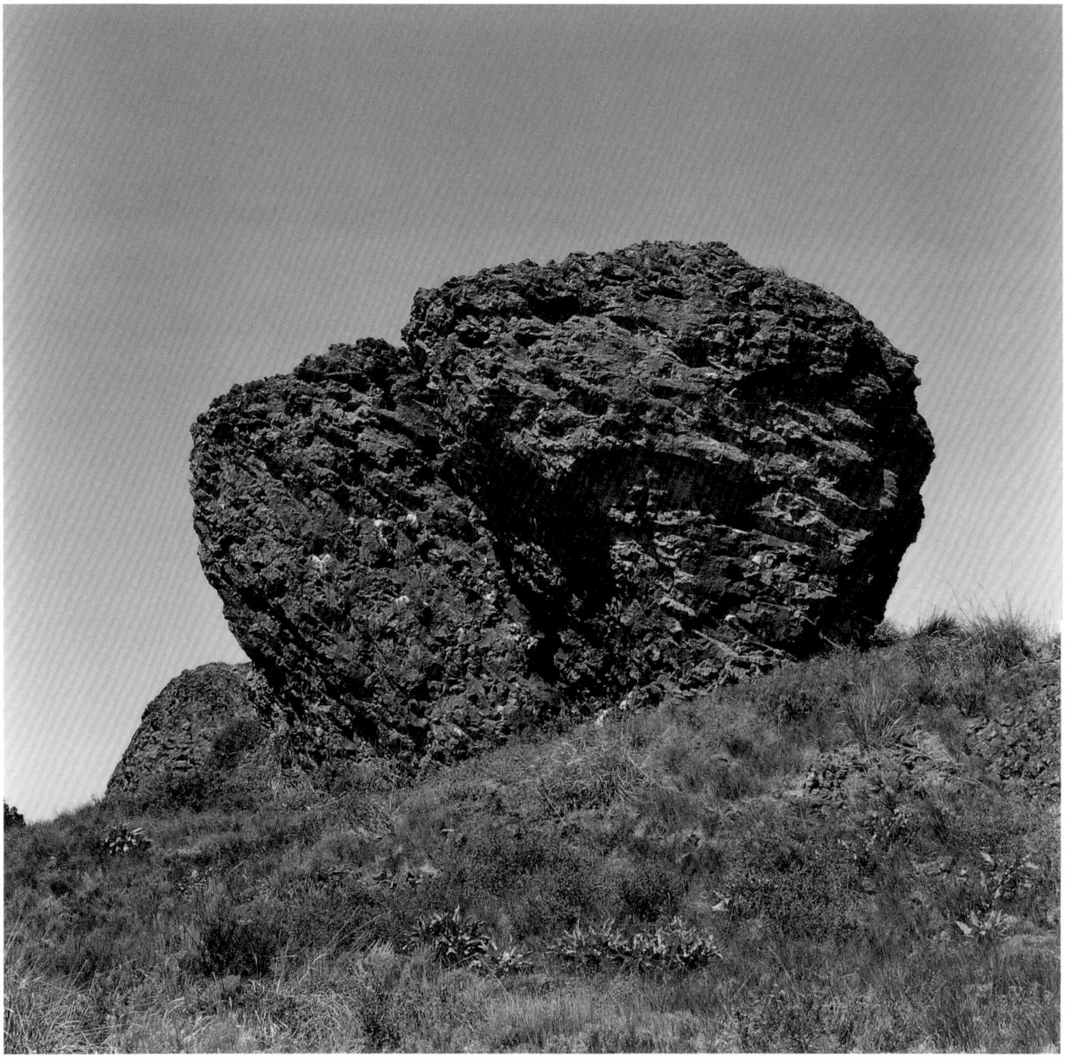

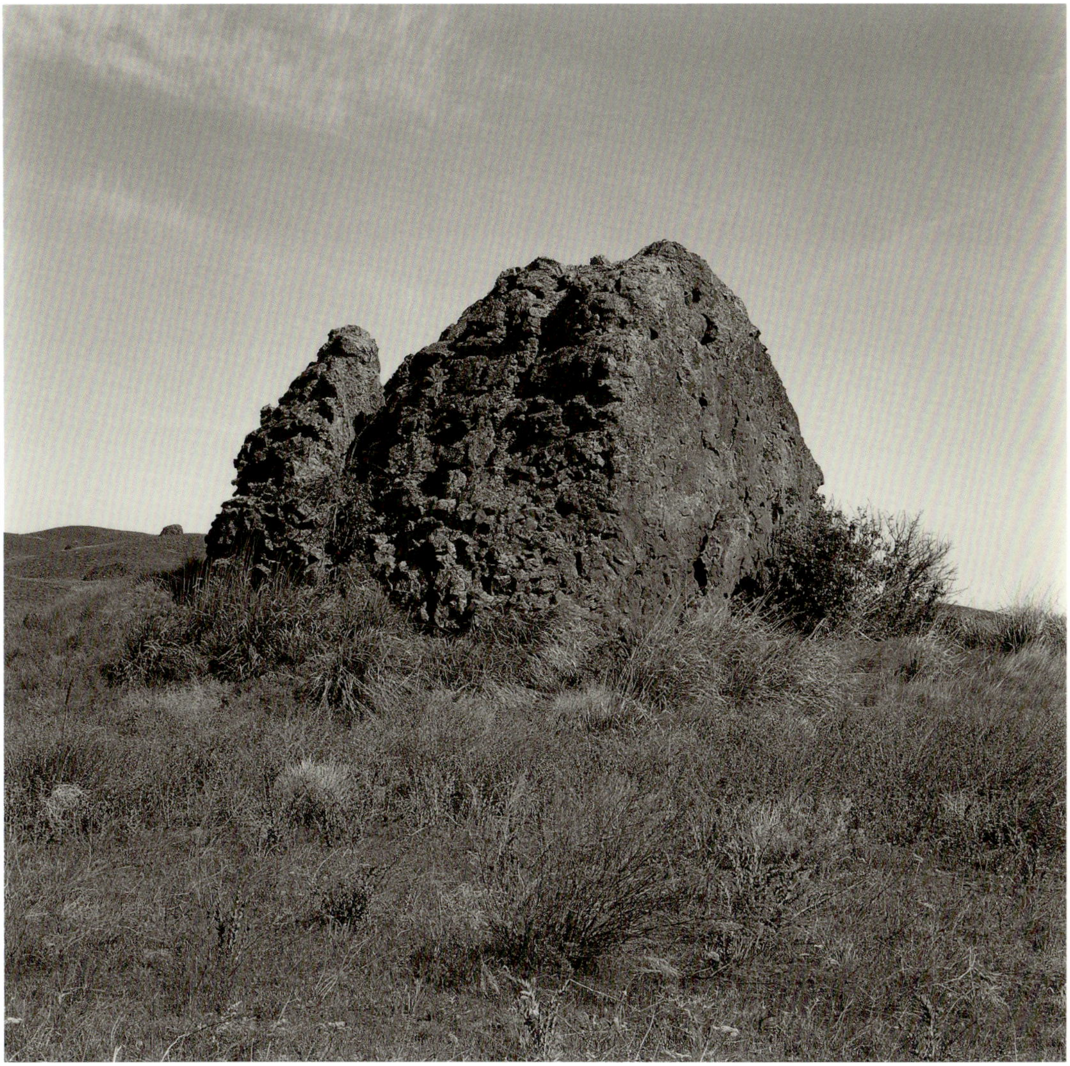

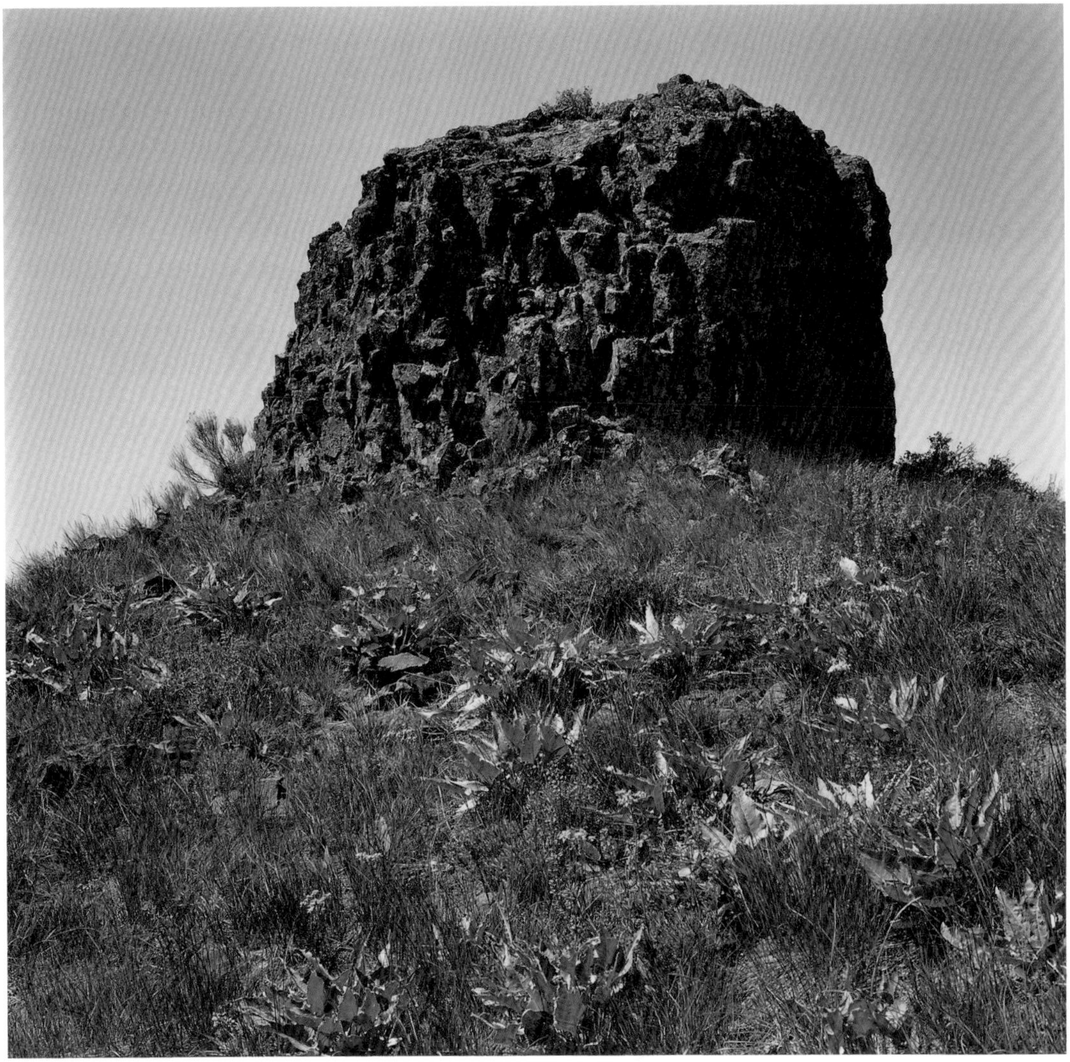

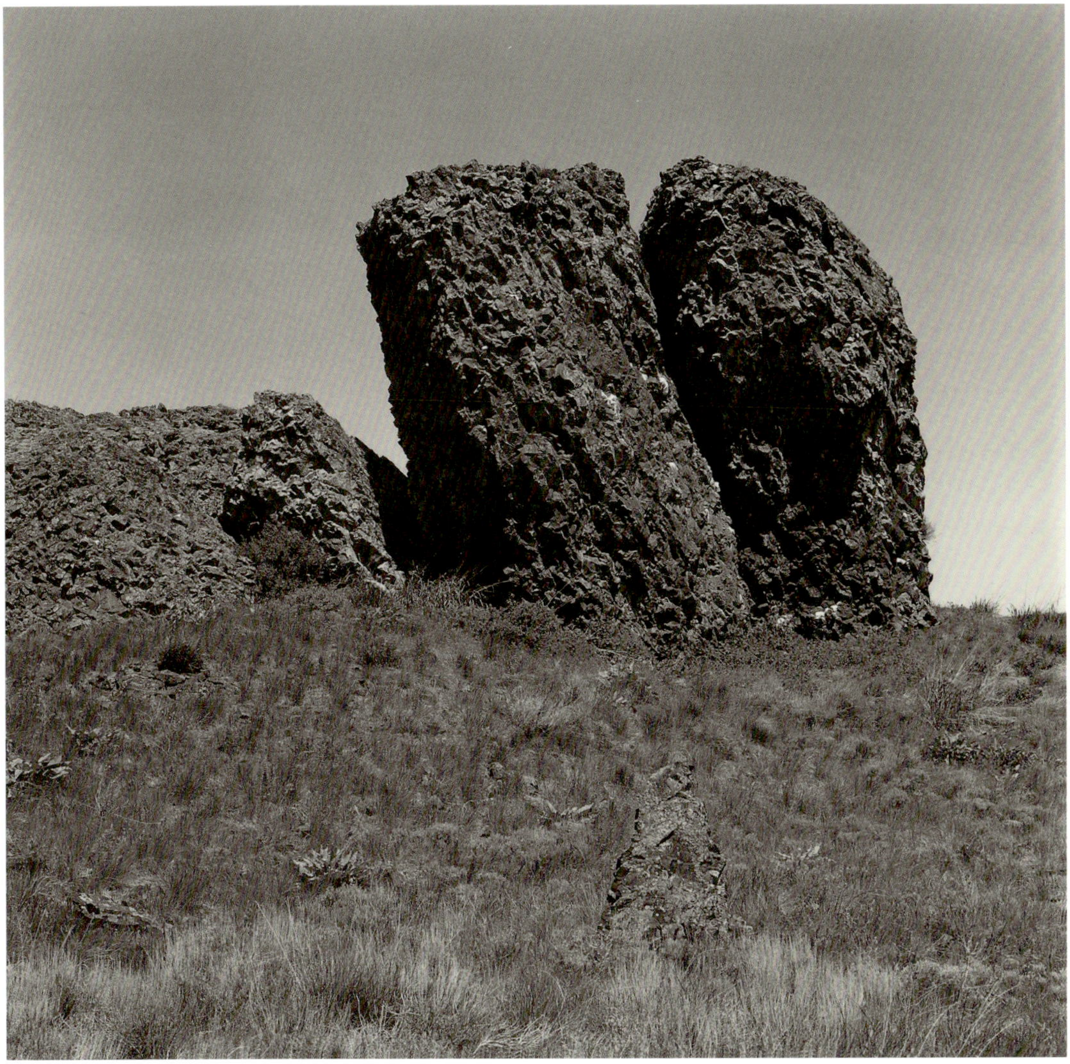

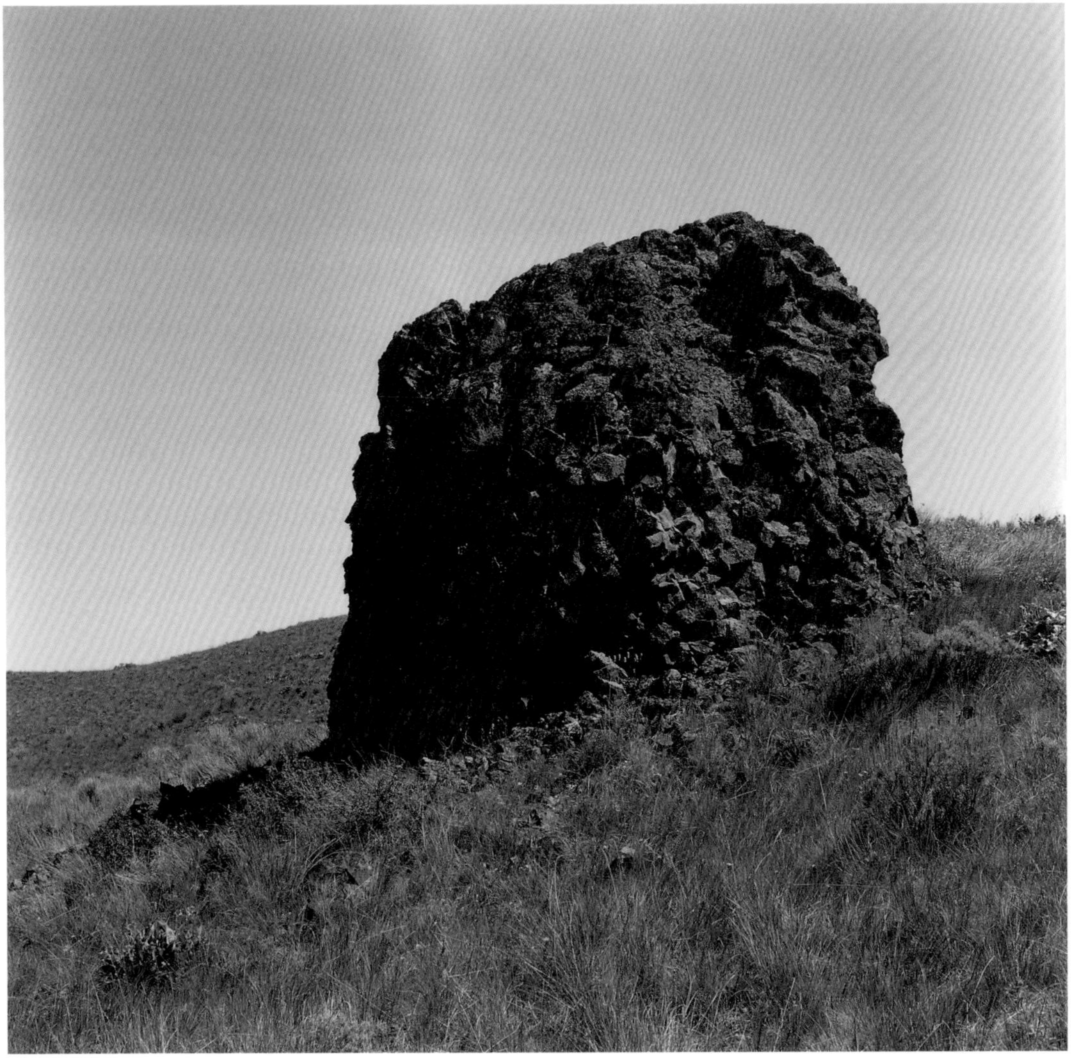

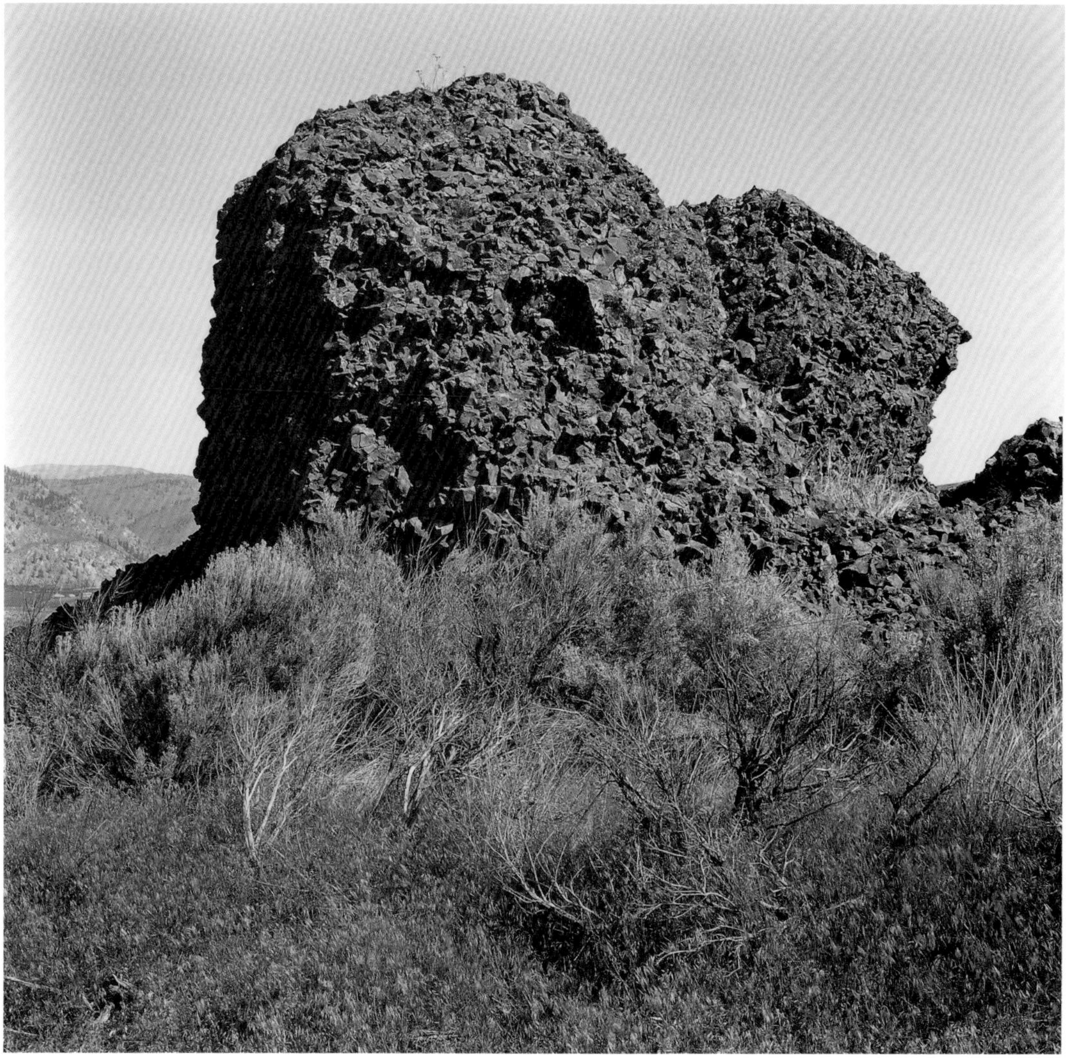

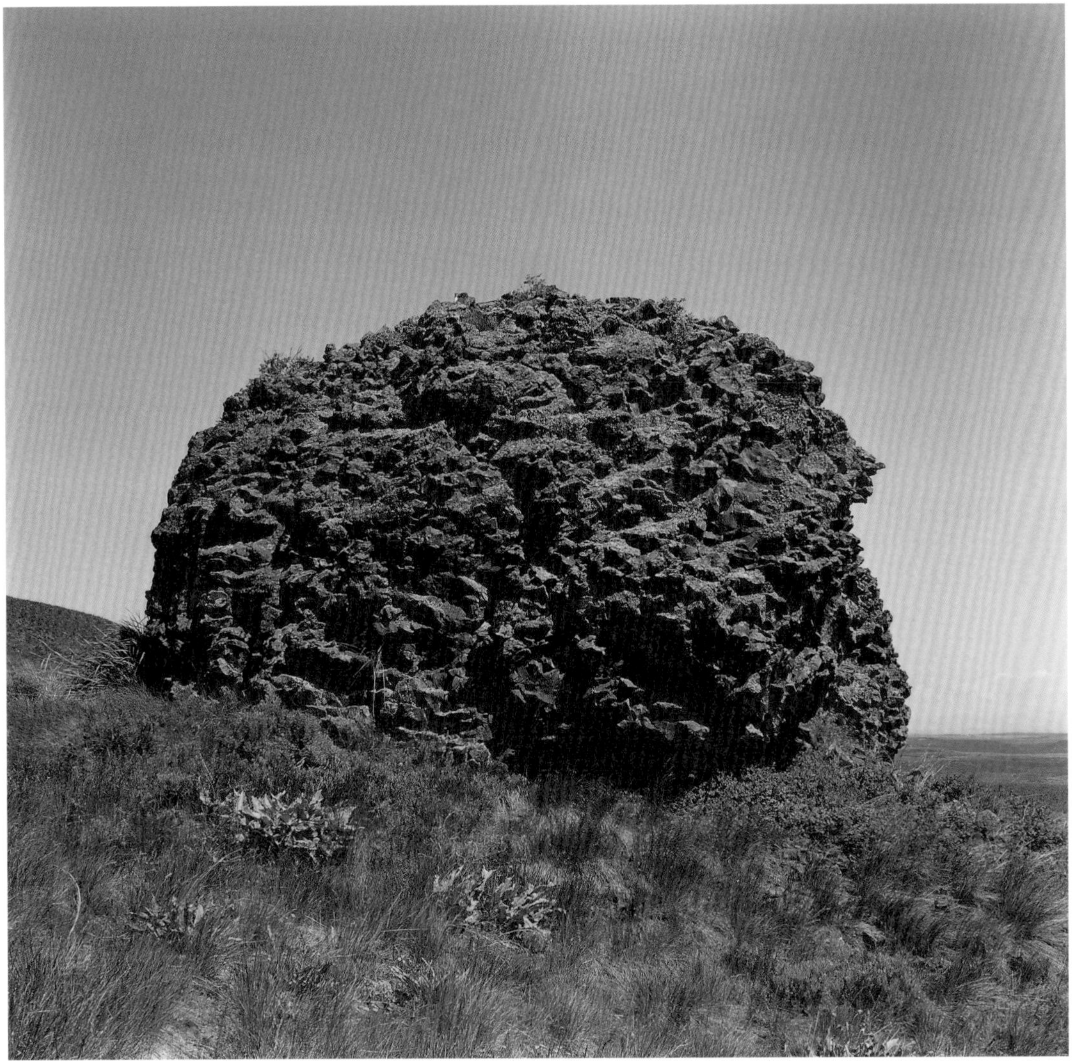

ACKNOWLEDGEMENTS

Li Po's poem "Zazen on Ching-t'ing Mountain" is from *Crossing the Yellow River: Three Hundred Poems from the Chinese*, translated and introduced by Sam Hamill (Rochester, NY: BOA Editions, Ltd., 2000). For more information on the Missoula floods and geologic evolution of the Columbia plateau, consult *Cataclysms on the Columbia: A Layman's Guide to the Features Produced by the Catastrophic Bretz Floods in the Pacific Northwest* by John Eliot Allen, Marjorie Burns, and Sam C. Sargent (Portland, OR: Timber Press, 1986).

Special thanks to Ian Boyden, Jennifer Boyden, Frish Brandt, Colleen R. Chartier, Gregory A. Hicks, Sarah Katz O'Hara, Beth Sellars, and Claudia Vernia, and with appreciation for the support of Roberta R. Katz and Sydney M. Katz.

COLOPHON

Erratics was published in conjunction with the exhibition "Erratics: Photographs by Charles Katz, Jr." at the Sheehan Gallery, Whitman College, Walla Walla, Washington (October 25–December 13, 2002). All type in this book is Perpetua, a typeface designed by Eric Gill (1882-1940). Gill designed the letters for Perpetua with the intention to convey the nobility, dignity, and monumentality of the letter forms he engraved in stone. The photographs are reproduced as 200-line screen duotones and printed on acid-free, recycled Centura silk paper with soy-based ink. The endpapers are Mohawk Ultrafelt. The book was printed by Watermark Press, San Francisco, California. The book is Smythe-sewn and bound by Roswell Bindery, Phoenix, Arizona. Book design by Ian Boyden.